the
end of
the
world

thomas y. crowell company   new york

the end of the world

franklyn m. branley
illustrated by david palladini

By the Author
THE CHRISTMAS SKY
THE MYSTERY OF STONEHENGE
MAN IN SPACE TO THE MOON
PIECES OF ANOTHER WORLD:
   The Story of Moon Rocks
THE END OF THE WORLD

Copyright © 1974 by Franklyn M. Branley
Illustrations copyright © 1974 by David Palladini

All rights reserved. Except for use in a review, the reproduction or utilization of this work in any form or by any electronic, mechanical, or other means, now known or hereafter invented, including xerography, photocopying, and recording, and in any information storage and retrieval system is forbidden without the written permission of the publisher. Published simultaneously in Canada by Fitzhenry & Whiteside Limited, Toronto.

Manufactured in the United States of America

Library of Congress Cataloging in Publication Data
Branley, Franklyn Mansfield, 1915-
  The end of the world.
  SUMMARY: A scientific explanation of how, millions of years from now, the life of the Earth could conceivably end.
  1. End of the world (Astronomy)—Juv. lit. [1. End of the world (Astronomy)] I. Palladini, David, illus. II. Title.
QB638.8.B7   525'.01   72-13264
ISBN 0-690-26607-3
ISBN 0-690-26608-1 (lib. bdg.)

10   9   8   7   6   5   4   3   2   1

16-WASHINGTON HIGHLANDS

Juv
525.01
5.50

To

T. E. B.
D. F. B.
K. E. D.
S. K. D.

Someday the world will end.
 This lush, blue-green planet will become a frozen, dark cinder. Or it may be evaporated into space. It may be broken into a multitude of smaller worlds—a family of small planets, hurtling into the depths of a dead, black sun, a sun altogether different from the sun we know today.

No matter how you look at it, the final outlook for man and his world is dismal. In his folly man may ruin parts of the earth, but the works of man and the destruction he causes are insignificant. Eventually the entire earth will be destroyed; mountains will be leveled; seas will disappear—all this will be done by cosmic forces over which man has no control, and which are as inevitable as tomorrow.

This is a gloomy forecast but one about which we should not despair. There are consoling factors. The end of the world will not occur in your lifetime, or that of your children's children. We can be quite sure that doomsday will be billions of years in the future. By that time man may have left this blue world of water to seek out a new homeland somewhere else in the galaxy—somewhere far beyond the domain of the sun.

Although scientists can give strong arguments that the world will end, no one can prove how it will happen, for no one has ever seen the end of a planet anywhere in the universe. This is mainly because we have never been able to observe any planets other than those in our own solar system. But we can observe stars, and we have seen some of them explode. If there were planets associated with such stars, the planets would have been heated intensely, becoming so hot that all life upon them would have been destroyed. The planets would very likely have been reduced to the gases from which they were made at the time of creation. As far as we know, the planets of our own solar system have all been here since the solar system came into being. No dramatic, planet-wide changes have been reported on any of them. Indeed, astronomers believe that the planets have existed in their present condition for a long, long time.

Most of the conditions that prevail on the planets, including earth, exist only because of the sun, the star which is the center of the solar system. The stars, including our sun, are not eternal, as poets tell us. Quite the contrary, stars are ever-changing—the one characteristic all stars have in common is change. Let's see how the sun has changed, how we expect it may change in the future, and how those changes will affect our planet.

The sun is a great ball of gases, mostly hydrogen, radiating stupendous amounts of energy. It supplies earth with energy which keeps us alive and does the work of the world. Our gas, oil, electricity, gasoline, and all of our food can be traced to the sun. This adds up to an incredible total, yet the earth receives only one two-billionth of the sun's energy. Nuclear reactions in the sun create this energy. In order to generate energy through sustained nuclear reactions, a star must contain enough material—hydrogen and other gases—packed together so that its internal temperature reaches several million degrees. The temperature starts a nuclear reaction. Once started, the reaction keeps going as long as there is fuel to feed it. Each second some four million tons of hydrogen are consumed in the sun's nuclear furnace. One would think that such a staggering amount would have used so much hydrogen that the sun could no longer produce energy. But the sun contains so much hydrogen that it will

release energy at its present rate for at least an additional five billion years.

No one knows how the sun came into existence. One theory, which scientists call the Big Bang theory, describes the universe's beginning as a tremendous explosion. All of the material of the entire universe, the scientists think, was once contained in one stupendous, densely packed, primeval "atom." (Where all the material came from and how it could be contained in this fashion are not clear.) Tremendous internal pressures caused this "atom" to explode. In seconds it ceased to exist. The material it had contained sped into space in all directions. Through eons of time the gases released from this "atom" continued to expand, and they still are expanding.

The gases moved at random; occasionally one atom of gas collided with another. When their speeds and angles of collision were just right, two freely moving atoms joined together. The mass of this new unit was greater than that of the two single particles, and its gravitational attraction was then great enough to attract other atoms. Through countless ages of time, atoms were added to atoms. This process of gaining mass is called accretion. The disorganized, randomly moving gases gradually joined together into a system. The more massive the system became, the more it was able to attract additional material. As the mass of the system increased, the packing together of atoms caused the temperature to rise to a million degrees, two million, and even more. Eventually the temperature became so high that hydrogen atoms (actually hydrogen nuclei, or protons) began to join together; nuclear reactions began and continued to occur. The gases had become a star.

For billions of years the sun has been radiating energy—keeping this world of ours at a nearly even temperature, providing the conditions which made it possible for life to emerge and develop.

During its lifetime the sun's activity has remained essentially steady. But the next illustration shows the slight variations in the release of solar energy that many scientists believe have occurred during the past five billion years. These changes, slight as they might have been, would have affected the earth profoundly, probably causing the ice ages. A decrease of only eight or nine percent in solar radiation causes the world temperature to drop low enough to prevent melting of the polar snow that falls during a given season. If the snow does not melt, it combines with the snowfall of the following season. Season after season the snow accumulates. After a few hundred or a few thousand years, glaciers form and move from the arctic regions into the temperate zones.

Early in the history of the earth, billions of years ago, very little of the sun's energy could reach earth because of heavy clouds that covered the planet. The earth was covered by deep layers of ice. Solar radiation increased gradually and steadily. Earth temperatures became sufficient to melt the ice blanket, and the warmth needed for the start of living organisms was present.

From time to time the regular and steady increase of solar activity was interrupted. There were abrupt but short-lived drops, resulting in falling temperatures on the earth. Some scientists believe that these slumps resulted from the decreased activity on the sun as indicated in the illustration. Others do not accept the explanation. They think that the drop in temperature was caused by conditions on the earth, and not by changes in the sun. They say that tremendous shifts in large segments of the earth's crust would have resulted in extensive volcanic activity. Heavy gases, especially carbon dioxide, thrown out by volcanoes would have surrounded the earth. These gases would have served as a "mirror" reflecting solar radiation back into space and reducing the amount of energy that reached the surface of the earth.

Scientists are not agreed on the reasons for these sharp drops in temperature, but whatever the causes, the changes could occur at any time, and it's conceivable that they can be more severe in the future than any the earth has experienced in the past. Our planet is now in the Quaternary ice age and the world temperature, that is, the temperature of the earth as it would be measured from a distant planet, stands at 57°F. Scientists believe that the temperature of the planet has increased steadily over the past few million years, and they compute that our world temperature should now be 72°. It is conceivable that whatever cooling conditions have kept earth's temperature from reaching the expected 72° might continue for a long, long time.

If this were to happen, glaciers would continue to grow. Each winter would bring its new snow falling on old snow that never melted and that was packed together to make a hard icy layer. Century after century the snow would pack and harden. For a thousand years the temperature would hold steady or drop only a little bit. Perhaps the drop would be no more than a degree in a million years. Small as this is, it would be enough to cause glaciers to grow. Cold polar winds would blow over the ice fields. Great air masses would carry cold air into the temperate regions of the earth. Slowly, growing seasons would become shorter—only an hour or so at first, but then a day, two days. As eons went by, patterns of life would change. People would grow crops that could endure colder weather. They would move toward the equator to find places where the weather was warmer, and existence less demanding. Populations would become denser, as the part of the earth where people could live became smaller. The struggle for existence would become more and more competitive—there would be less living space, less farmland for growing food. The minerals and ores found in the temperate zones, far removed from the equator, would be buried beneath rivers of ice.

For several million years the glaciers might grow and advance toward the torrid zone. Epidemics or wars might reduce the closely packed population of the world. Existence would be difficult, but all life would not disappear; mankind would survive.

Judging by what we believe about the performance of the

sun and other sun-like stars during the past five billion years, solar activity would increase eventually. The temperature of the earth would slowly begin to rise. Instead of pushing toward the equator, the leading edges of the glaciers would melt slowly. As they did so, new ice would be pushed along. After untold ages of time, the melting would proceed faster than new ice was formed. The glaciers would begin to recede. Winter snows would again melt away in a single season. The ice and snow of millions of years would slowly melt, leaving behind lake beds, deeply gouged valleys, mountains worn smooth by the grinding action of tons of ice laden with rock, sand, and gravel.

As the glaciers receded and the earth became warmer, people would be able to move away from the tropical areas and back into the temperate regions of the earth. Continents that had been covered by ice and snow would once again become alive with plants, animals, and humans.

During the eons of its existence, the climate of the earth has gone through many such changes. Ice ages lasting millions of years have come and gone, yet life has persisted. It has flourished after each calamity, returning to the regions scarred by the glaciers. Such world-wide changes on earth disrupt life, but eventually the planet becomes as full of living things as it ever was. However, it is possible that the earth will be changed eventually in ways that reduce life permanently; masses of people might be destroyed and the earth so altered that life could no longer thrive upon it. Let's take a look at some of the possibilities.

The moon might be the cause of one of these disasters. Some astronomers believe that eventually, the moon will be destroyed. Should this happen, the destruction would cause drastic changes on the earth.

At this moment the moon is producing tides on the earth. In some places the change from high tide to low tide is only a few inches; in other locations it is thirty feet. But in the long past history of the moon, long before geologic periods now recognized, scientists believe that tides were eight thousand times greater than they are now. The variations from high tide to low tide were as great as the depths of the oceans.

It is believed that shortly after the earth and moon were formed, the moon was only 12,000 miles away from the earth. (Its present distance is about 240,000 miles.) During the earlier period the earth rotated very fast. A day on earth, the time needed to complete one rotation, was only about 4.8 hours. Not only was the earth spinning very fast, but the moon was also speeding around the earth. Today the moon takes $27 1/3$ days to go around the earth; in the early days of creation the moon revolved around the earth in much less time, only about 7 hours.

Ever since the moon was formed, it has been producing tides on the earth—both sea tides and land tides. (You are aware of sea tides, but the moon also affects the land, causing it to rise and fall six inches in some places.) The pull of the moon tends to slow down the rotation of the earth— the tides have a braking action. The effect is not very great, only enough to lengthen the day about one second in one

hundred thousand years. But the braking action occurs year in, year out. While the earth's rotation period lengthens, the moon moves farther from the earth. Each year the moon is about 1.5 inches farther from us than it was the year before.

If we project billions of years into the future, we can expect that the day will continue to get longer. And so will the month, for as the earth-moon distance increases, the time required for the moon to go around the earth becomes greater. Eventually the day and the month will equal one another; both of them will be about fifty 24-hour days long. The same half of the earth will always be facing the moon. If you happen to be on the half which is away from the moon, you would never see the moon, except by traveling to the other side of the earth. By this time people's lives would probably no longer be regulated by the day-night cycle that we know today.

The moon will have moved so far from the earth that its tide-causing gravitation will have become very small. The gravitation of the sun will be the principal tide-causing force. Tides caused by the sun will slow earth rotation even more. Slowly the day will become longer than the month. This will cause the moon to move in closer. Once more the earth will speed up. The energy in the earth-moon system remains the same; if the moon moves closer, the earth speeds up, if the moon moves farther away, the earth slows down. The distance between the earth and the moon will not change rapidly—only an inch or so a year, but year after year, they will come closer and closer. As the moon moves closer to the earth, tides will become higher. For thousands of years the change will not be apparent. But a time will come when structures at or close to sea level would be flooded. As time passed, the water would reach higher. Whole coastal cities would be flooded. People would move inland, establishing colonies on high mountains. Even there they would barely be above the rising tide. Flooding would reduce the amount of land for growing crops; famine and epidemics would sharpen the struggle to survive; and large parts of the population of the world would be destroyed.

The tides would be catastrophic. As each high tide advanced, the water would be a towering wall sweeping away everything in its path. Waters that had been piled a mile high would crash and pound back into the hollows of the oceans. Great whales and sharks, should they still exist, and fishes of all sizes would be trapped, stranded out of water and not able to live until the next wave of water covered the land. As the moon moved closer and closer, the days would begin to grow shorter. The tides would become higher and higher, spreading destruction across entire continents. Then, all of a sudden, the moon would disappear in one great shattering explosion. There would be no more moon and no more tides.

The moon would shatter because it would have reached what scientists call Roche's Limit. Edouard Roche was a French astronomer who around 1850 computed that a satellite of a planet has to be at a distance from the planet which is more than 2.44 times the radius of the planet. If a satellite

crosses that critical danger zone the difference in gravitational force between the near side of the satellite and the far side would be enough to shatter the satellite. Some of the debris would rain down on the mother planet. But most of it would form into planet-circling rings. The theory of Roche's Limit seems to explain the existence of the rings of Saturn. We know the rings are made of separate particles, and it has long been suspected that these particles are the remnants of a satellite that was ripped apart by the mother planet. The outer edge of Saturn's ring system is at a distance 2.30 times the radius of Saturn—well within the danger zone. Mimas, the nearest satellite beyond the rings, is at a distance 3.11 times Saturn's radius. This places it well beyond Roche's Limit.

It is possible that at some time the moon might reach the danger zone—2.44 times earth's radius, or a bit under ten thousand miles. The moon could not exist once it entered the zone. It would be shattered. The small pieces of the moon would achieve their own orbits around the earth, forming a broad ring system. There would be no more great tides, only those caused by the sun. The rings would not be solid; they would be made of particles with spaces between them. But depending upon their density, they would cast shadows on the earth, reducing considerably the amount of solar radiation the earth would receive. During half its journey around the sun, the earth would receive no direct sunlight in the region shadowed by the rings. If this happened, crops could not be grown; ice and snow would accumulate; tremendous wide-spreading glaciers would develop, ushering in an endless ice age. It would be without end, because the region would receive direct sunlight for only a brief part of the year. Snows would not melt; they would continue to fall and to accumulate.

Areas of the earth now populated would become barren. Their climate would become too rigorous for the survival of plant or animal life. People who had survived the stupendous tides caused by the nearby moon would migrate across the barren earth, seeking those rare places where temperatures would be moderate and where sunlight and rainfall would be sufficient to support the crops essential to man's existence. Although man could not live widely over the earth, he could exist in limited regions.

Sudden decreases in solar activity might cause disastrous glaciers, or dense rings of lunar remnants might shadow extensive regions of the earth with the same final result. In either case, life would not be destroyed completely. Plants and animals would continue, but they would not flourish as they do today.

However, ultimately, life will disappear completely. This warm planet will lose all its heat, and the sun will no longer supply it with energy. The earth will be doomed—it will become a cold, barren world. Seas, lakes, oceans will be frozen solid. Moisture in the atmosphere will be squeezed out. It will fall to earth forming a mantle of snow and ice. Even the atmosphere will be frozen solid. For eons of time, life will be suspended on a world hurtling through space. But

chances are that man will not have survived long enough to experience this frigid finale. The sun will probably end as a cold star. Before it becomes a cold object, however, the sun will become larger and hotter, so hot and so large that earth, and other planets too, will be burned to cinders.

The sun has been shining for billions of years, varying ever so slightly in its steady production of energy. But like all stars, the sun is not everlasting. It has a life history to follow—birth, youth, maturity, middle age, old age, and death.

At the present time, the sun is probably a middle-aged star. For billions of years it has been ejecting energy at a rate that has been increasing slowly. There have been moments when the output has dropped. And there have probably been occasions when the output has momentarily risen steeply. But, on the whole, energy output has risen slowly.

During the lunar walks of Armstrong and Aldrin on the Apollo 11 mission, three-dimensional photographs of small three-inch-square areas of the moon were taken with stereoscopic cameras. Close inspection of the pictures revealed that parts of the moon are covered with tiny craters—some only a fraction of an inch across. Curiously, small, fragile columns of lunar dust were often found inside these craterlets. The appearance of the tops of the columns, and some of the walls of the small craters, fascinated the investigators. The top surfaces of the columns were glazed and smooth. So were certain walls of the craterlets.

After careful study of the formations, the most reasonable explanation was that the surfaces had been melted. This could have occurred only if there had been intense heating for a brief time. The most probable source of such flash heating was the sun. Many scientists believe that at some stage in the history of the moon, energy from the sun increased many times over its normal output. The energy swept the moon as a wave might wash across a beach. It was momentary, intense enough to cause the melting but so short-lived that the energy could not penetrate to any appreciable depth.

Such a sudden burst cannot be explained as a normal process of a star. But it is conceivable that a comet might have been pulled into the sun by its tremendous gravitational attraction. If this happened, the material of the comet would have been converted into energy instantaneously. A flash of tremendous heat and light would result.

If the moon were swept by such a wave of intense energy, so also might have been the earth. If such a calamity occurred in the past, it could also happen in the future. Fortunately, earth is surrounded by an atmosphere containing gases, dust particles, and water droplets, all of which (the latter two especially) serve effectively as insulators. However, if the upsurge of energy were to last any time at all, it would be sufficient to penetrate the atmosphere. If this were to happen, all vegetation would be burned completely in seconds, paint would be blistered and peeled from surfaces, the flash-point of wood and paper would be reached. The earth would be set afire. The only safe places would be caves

and the waters of the earth. Surface water might be heated tremendously, but beneath the surface it would hold to its constant temperature of about 55° F.

Such a flare-up of the sun would be cataclysmic; it would devastate the countryside, causing widespread famine and death. But this alone would not destroy mankind. Not all creatures would be killed; some would survive to establish the species once again.

The outlook is more final when one considers the normal evolution of the sun. No one has ever observed the life history of any star. However, different stars at different stages have been studied. When astronomers put together their observations, they are able to surmise that the life history of the sun may proceed somewhat as follows.

The sun is now in a state of equilibrium—it is a balanced star. Energy produced in the solar interior exerts an outward pressure. This is more than enough to cause the sun to expand to many times its present size. But the sun is massive (more than 300,000 times more massive than the earth), and so it has a strong gravitational field. The sun's gravitation tends to collapse the sun, make it smaller. But the energy produced in the interior tends to expand the sun, make it larger. Contraction and expansion are balanced and so this star of ours maintains its volume, and has been doing so for several billion years.

At present the sun is a hydrogen-burning star. Its hydrogen nuclei are fusing to produce helium nuclei. The nuclear process results in a decrease in the amount of hydrogen nuclei and an increase in helium nuclei. Eventually, after some seven or eight billion years, the proton concentration will drop so low that the fusion process cannot continue. The sun will begin to cool. Gravitation inward, no longer balanced by radiation pressure outward, will cause the sun to grow smaller. As the sun contracts, the material of the sun will be packed together so tightly that the sun's temperature will increase rapidly—this time because of contraction rather than nuclear activity. The interior solar temperature of fourteen million degrees may reach one hundred million degrees.

The high interior temperature results in tremendous outward pressure. The outer regions of the sun will expand explosively; perhaps the sun's diameter will expand from about a million miles (its present diameter) to thirty million miles.

As the sun becomes warmer, so will the earth. Our present planetary temperature of 57° will reach 60°, 70°, 80°, and keep rising. Should there still be people, they will migrate out of the tropics and toward the polar regions. At first this will solve the problem. But temperatures will continue to rise. The oceans will get warmer, and the great currents which now result from temperature differentials will cease to flow. Fishes and other sea life that require a low temperature will perish. Caves will provide temporary protection from the heat. Seedlings will be burned by the searing sun. When food reserves are consumed, all life will disappear. The earth itself will become unbearably hot.

Water will evaporate into the atmosphere, and at first it will rain down upon the earth. But the rain will be hot. After millions of years, there will be no more rain. The seas will have boiled away; the atmosphere will have become so hot that water and other particles, the atmosphere itself, will have accelerated into space, leaving a barren world.

All the ice and snow of the earth will have melted long ago. The few people who have survived, if any, will find that even deep caves provide little refuge, assuming they have a supply of oxygen to sustain them. The earth will be a seared ash; all water and air will be boiled away. Our planet will be a dead, inert world, no more alive than the moon is today, and no more able to support life in any form.

What about man? Will this be the end of him? It surely will be, as far as planet earth is concerned. But long before that time, his technology may have developed far enough for him to set out upon journeys into deep outer space, journeys that might take him to another star—to another planet with air and water.

At this time we do not know of any such planets. We have not actually seen any planets other than those which belong to our own solar system. But we do have observations of certain stars that indicate the presence of planets. In the constellation Ophiuchus there is a star called Barnard's Star, which is named after Edwin E. Barnard, the American astronomer who first observed it in 1916.

Thousands of photographs of the star have been taken

since it was discovered. These photographs show that Barnard's Star does not move in a straight line through space as a solitary star should. It has a side-to-side motion, more than any other star that has been investigated. Dr. Peter van de Kamp and his colleagues at Swarthmore College undertook a careful analysis of the photographs. Their study indicated that the wavy movement of Barnard's Star was caused by the movement around the star of a mass 1.5 times the mass of Jupiter. This means that there is a planet associated with Barnard's Star. Of this we can be quite sure. Whether the planet supports life, or could become a haven for man in his flight from a doomed earth, is quite another matter. Compared to other stars, Barnard's Star is close to earth, only 5.9 light years away; Alpha Centauri is the only star system that is closer. A spaceship moving one hundred thousand miles per hour would take some fifty thousand years to get to Barnard's Star. Obviously, such a ship would have to be self-sustaining and have a closed biological system that could support itself for generations by the recycling of materials. This is not a farfetched idea, for earth itself is such a spaceship—one that has just so much material to work with and which can become depleted through unwise use of its resources.

Undoubtedly there are other stars that have planets associated with them, but they are at greater distances from earth. A colony of men and women equipped to journey through space for thousands of years might find a haven.

If any people could get away from earth, their ship would have to move away from the sun or what was left of it, out of the solar system, and into the realm of another star. For after the sun had grown into a giant star and cooled to a red star extending halfway to Mercury, it would pass through a series of events, each of them momentous, and all leading to its demise.

The accumulation of helium in the core of the sun, and the temperature increase to a hundred million degrees and more, would once again make the sun into a nuclear furnace. But hydrogen would no longer be the fuel. You recall that hydrogen was converted into helium. Now the helium would become the fuel. Helium nuclei would combine in a series of steps that would produce the nuclei of carbon atoms. Heavier elements would be produced as well. Eventually the amount of helium would drop so low that nuclear reactions would slow down. Because less energy was being produced, the temperature would drop. The gravitational attraction would now be stronger than the outward pressure, and the sun would collapse.

According to some theories, as materials packed tighter and tighter, the temperature would go up again. As it became higher, nuclear reactions resulting in still heavier nuclei would occur. Steadily the temperature would rise, perhaps reaching five billion degrees. Because of this high temperature, the sun would expand explosively, maybe becoming a sphere with a diameter of three hundred million miles. Earth would be inside the sun. The high temperatures would vaporize the earth and it would be entirely lost within the solar mass.

Eventually the pressures would be so great that perhaps the sun would blow up, throwing off vast amounts of gases. Some of the gases would be pulled back in, but others would be lost forever.

Material remaining in the sun would cool. Matter within it would pack together tightly, since gravitation would be much stronger than the pressure pushing outward. The massive sun, today a million miles across, will have shrunk to a millionth its former size—no larger than earth is now. Nuclear reactions will have ended, and all heat will come from contraction.

The sun will have become a white dwarf, one in which matter is so concentrated that a cubic inch will weigh one hundred tons. It will be a small, insignificant star, unnoticed among its brilliant neighbors.

Some astronomers believe there is one step beyond this that may apply to stars such as the sun. The star cools even further and all color fades—the white dwarf becomes a black dwarf. It contracts so much that its density approaches infinity. The gravitational field becomes so strong that all masses are pulled into it. Indeed, gravitation is so powerful that even light and all other radiation are pulled into the star. The black dwarf would not be seen. No astronomical instruments, not even radio telescopes, could detect this black dwarf star. It would be a black hole in space, pulling the remains of the solar system into its depths.

The sun and its planets, their satellites, the comets and all other parts of the solar system would have spun out their existence.

No doubt other planetary systems in the far reaches of space are ending their life cycles at this moment and perhaps in this very fashion. And no doubt every day of our lives other cycles begin in our own galaxy and in the galaxies far beyond.

New stars are born. New planets gradually come into existence. Life goes on. Not on planet earth, perhaps, but at locations still unknown to humankind.

ABOUT THE AUTHOR

Franklyn M. Branley, Astronomer Emeritus and former Chairman of The American Museum-Hayden Planetarium, is the author of many books, pamphlets, and articles on various aspects of science for young readers. He is also co-editor of the Let's-Read-and-Find-Out science series.
    Dr. Branley holds degrees from New York University, Columbia University, and the State University of New York College at New Paltz. He and his wife live in Woodcliff Lake, New Jersey, and spend their summers at Sag Harbor, New York.

ABOUT THE ILLUSTRATOR

David Palladini was born in Italy, but came to the United States when he was very young and grew up in Highland Park, Illinois. He received his art training at Pratt Institute in Brooklyn, New York. In addition to illustrating books, Mr. Palladini has received many awards and citations for his work in poster design and the graphic arts.

DISTRICT OF COLUMBIA PUBLIC LIBRARY

DATE DUE	DATE DUE
OCT 7	
JAN 11 1984	NOV 6
	DEC 28 1983

Juv   16-WASHINGTON HIGHLANDS

**The Public Library
of the
District of Columbia
Washington, D.C.**

P.L. 117 revised

Theft or mutilation
is punishable by law